Caring for Wildlife

By Belinda Gallagher

RUBY TUESDAY BOOKS

Published in 2024 by Ruby Tuesday Books Ltd.

Copyright © 2024 Ruby Tuesday Books Ltd.

All rights reserved. No part of this publication may be reproduced in whole or in part, stored in any retrieval system, or transmitted in any form or by any means, electronic, mechanical, photocopying, recording, or otherwise, without written permission from the publisher.

Editors: Ruth Owen & Mark J. Sachner
Production: John Lingham

Photo credits:
Alamy: 4T (Ernie Janes), 4C (Britpix), 4B (Simon Dack News), 10T (Ian Lamond), 12T (Kay Ringwood), 12C (Arterra Picture Library), 14B (Mike Lane), 17BR (EB Photography), 28T (Roger Parkes); Susie Moon: 26B; Nature Picture Library: 15CR (Laurie Campbell); Ruby Tuesday Books: 7R, 11B, 13R, 19B, 21B, 29T; Shutterstock: CoverTL (Ostanina Anna), cover CL (Midori Photography), cover BL (Madlen), cover TR, cover BR (Oksana Schmidt), 5TL (urbans), 5BL (MNStudio), 5TR (Colleen Ashley), 5BR (BernadetteB), 6T (Wild Media), 6CL (David Savile), 6CR (FJAH), 6B (DKW Images), 7T (Baulina Ekaterina), 7C (Melanie Hobson), 7B (W. de Vries), 8T (Ihi), 8CL (V Coscaron), 8CR (Sarah2), 8B (Christopher MacDonald), 9TL (Alla Simacheva), 9TR (Ivonne Wierink), 9BL (TrotzOlga), 9BR (aopsan), 10B (Jaroslav Moravcik), 11TL (taviphoto), 11TR (photographyfirm), 12BL (Caroline Blackburn), 12BR (Muafik89), 13TL (Maria Evseyeva), 13BL (Ihor Hvozdetskyi), 14T (TSN52), 14CL (Monkey Business Images), 15T (Imfoto), 15CL (xpixel), 15BL (Anest), 15BR (Yevhenii Chulovskyi), 16T (Tomasz Klejdysz), 16C (Henrik Larsson), 16B (Carl McKie), 17TL (iMarzi), 17BL (irin-K), 18T (HTU), 18C (Marina Green), 18B (Marina Demidiuk), 19T (Jamie.ff), 20T (Ciocan Daniel), 20CL (Ingrid Balabanova), 20CR (David Obrien), 20B (Cat28), 21T (Eileen Kumpf), 22T (Tatiana Buzmakova), 22B (Miodrag Zlatarov), 23TL (Reimar), 23TR (Andriy Blokhin), 23BL (PhillipsC), 23BR (Lindaze), 24T (Mark Bridger), 24C (Judy Kennamer), 24B (Dan4Earth), 25T (Serenity Images23), 25C (FJAH), 25B (Mikhail Novokreshchenov), 26T (stmilan), 27T (Astrid Gast), 27C (grayjay), 27B (Martin Fowler), 28CL (Ann Gaysorn), 28CR (aleks733), 28B (nieriss), 29B (Chamois Huntress), 30L (Kate Russell), 30TR (Muddy Knees), 30BR (Swapnil Anghan), 31TL (Will Howe), 31BL (Mr Har), 31TR (Robert Schneider), 31BR (palspicts).

ISBN 978-1-78856-443-4

Printed in Poland by L&C Printing Group

www.rubytuesdaybooks.com

Note from the Publisher

Neither the publisher nor the author can accept legal responsibility or liability for any loss, harm or injury that may come about from following the instructions in this book. All activities should be carried out with adult guidance and supervision. Some activities involve being out of doors in public spaces. Children should be accompanied at all times. It is the parent's or carer's responsibility to ensure their child is safe.

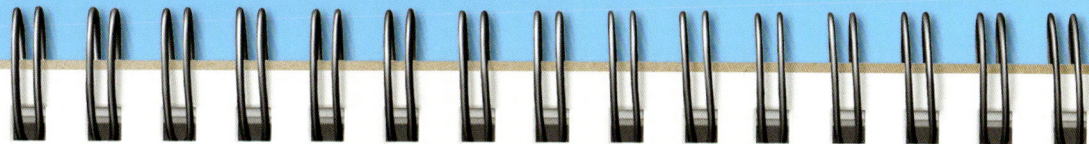

CONTENTS

Wildlife Needs Your Help 4

Caring for Habitats 6

Let's Clean Things Up! 8

Go Wild to Be Wildlife Friendly 10

Make Some Microhabitats 12

Minibeasts Are Cool and Damp 14

Beautiful Beetles 16

Create a Buzz! 18

Create a Butterfly Garden 20

Let's Feed the Birds 22

Best for Nests 24

Make a Splash for Water-Lovers 26

Keep Caring for Wildlife 28

Glossary 30

Index 32

Staying Safe!

All the activities in this book are fun and easy to do. Be sure to ask an adult to help you with each one at every stage. Never go anywhere without your trusted adult. Wear old clothes for the make-and-do activities. Always make sure an adult is nearby when using scissors or garden tools. When outside, make sure you are wearing clothes that suit the weather. Have fun!

Wildlife Needs Your Help

We are surrounded by wildlife. Birds, **mammals**, **insects** and **amphibians** live everywhere – but wild animals are struggling!

Their natural homes are becoming dirty and unsafe because people aren't taking care of the natural world.

Plastic milk bottle on a beach

LITTER is dangerous for wildlife. Animals may eat it or become trapped in it.

PLASTIC rubbish gets into rivers and washes out to sea. This plastic puts ocean animals in danger, and it washes up on beaches.

Wild places are destroyed to make space for houses, shops, factories and roads.

However, if everyone helps, wildlife can live in the smallest of spaces – even in a city!

Just one small plant in a pot can provide food and shelter for insects.

Strawberry plant

Wildlife Water Bowl

Birds and insects need fresh water. Make a place where our smallest wildlife can safely get a drink — without falling in!

You will need:
- A small shallow bowl, dish or other container
- Clean water
- Stones of different sizes
- Twigs and leaves

1. Put your container outside. Pile up stones or place them around the container to make rocky steps.

2. Fill the container with fresh water. Remember to top up the water regularly.

You can add leaves and twigs as extra places for bugs to balance and hang out.

Top Tip
You can prevent a wild animal from getting hurt or sick by picking up litter. Can you start today?

There is so much you can do to help wildlife. If you can create a wild space, there will be animals waiting to move in!

Caring for Habitats

The place where an animal lives, finds food and cares for its young is called a **habitat**. We must take care of these places and keep them litter-free so that wildlife can survive.

Fox cubs

When you visit a natural habitat:
- Don't disturb wildlife.
- Bring all your litter home.
- Pick up any litter you see.

KEEP BEACHES CLEAN

Seabirds, seals and other animals live on and around beaches. This habitat is home to nesting seabirds.

Ringed plover seabird

CARE FOR WOODLANDS

Chick

Great spotted woodpecker

Woodlands are packed with trees and other plants. Animals make their homes in tree trunks, mossy logs and thick bushes.

Harvest mouse

GIVE HEDGEROWS A HAND

Hedgerows provide food for birds and small mammals. Mice and birds make their nests in hedgerows.

MAKE TIME FOR MEADOWS

Peacock butterfly

Meadows are packed with wildflowers that provide insects with sugary **nectar** to feed on. Can you sow some wildflower seeds in your garden or school playground?

Top Tip
Plastic bags often get caught in hedgerows. Can you stop using plastic bags?

PICK UP FOR PONDS

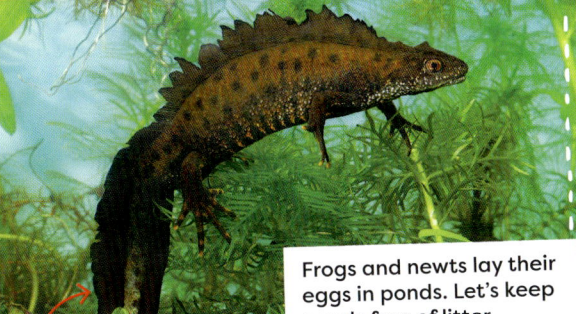

Frogs and newts lay their eggs in ponds. Let's keep ponds free of litter.

Great crested newt

Plant a Tree

Trees provide lots of different habitats for wildlife, from bark to leafy branches.

You will need:
- A potted sapling (young tree) from a garden centre
- Gloves
- A spade
- A watering can
- Peat-free potting compost

1. Choose a spot for your sapling where it will have enough room to grow.

2. Dig a hole twice as big as the sapling's pot.

3. Gently tip the sapling from its pot and place it in the hole. Soak the roots with water.

4. Cover the roots loosely with the dug-out soil. Press the soil down with your hands.

5. Add a layer of compost to the top of the soil. Give one more soaking of water.

Keep watch for wildlife visiting your tree!

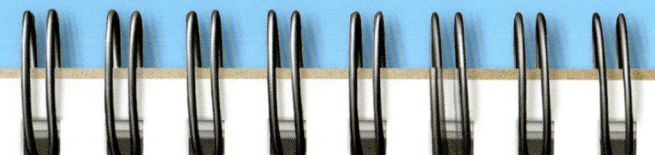

Let's Clean Things Up!

Litter is always bad. But if it's thrown away in natural habitats, it's a danger to wildlife. Let's look at how we can keep our world clean and safe.

A hedgehog trapped in a plastic can holder

Picking up other people's litter isn't nice. But if you see some litter, do your best to clean it up. Encourage friends to use bins and be responsible for their litter.

Always wash your hands as soon as possible after touching litter.

Say no to plastic bags, straws and single-use plastic cutlery, cups and plates. Once they are used, these throwaway things just become more rubbish.

Did you know?

Plastic items never disappear! If someone drops a plastic bag, spoon or cup in a wild place, it just breaks up into smaller and smaller pieces. It pollutes the habitat and may be eaten by an animal.

Try not to use single-use drink bottles and packaged takeaway food.

Reusable drink bottle

Reusable lunchbox

Top Tip
Pack a snack! On days out, instead of buying snacks, take your own. A homemade sandwich tastes better and comes in a reusable box.

Litter Pick and Sort

Gather some friends and clean up your street or an area where you live. The litter you collect can be sorted into groups for recycling.

You will need:
- Gloves
- Reusable bags

1. Work as a team to pick up litter and put it into your bags.

2. When you are finished, sort the litter into groups for recycling. Which group was the biggest?

Glass bottles and metal cans can be recycled and made into new bottles and cans, again and again. Some plastic items can also be recycled and made into new items. Ask an adult to help you sort and recycle any litter you collect.

If everyone picked up just a few pieces of litter, it would soon amount to A LOT!

9

Go Wild to Be Wildlife Friendly

Did you know that one easy way to help wildlife is to be messy? Wildlife needs gardens and outdoor spaces that are a little untidy.

Let plants grow wild in your untidy area.

Add small piles of wood, stones or old bricks to make an animal hideaway.

Learn to love **weeds**! Allow these wild plants to grow. The less you cut them back, the more wildlife will like it.

Let grass grow long. It provides shelter for small animals such as frogs.

Bee

Dandelion

Pollen

Top Tip

Sometimes people kill weeds by using chemicals called weedkillers. Ask the adults you know to NEVER spray or pour these poisonous chemicals onto their gardens.

Dandelions are pretty weeds, or wildflowers, that provide nectar and **pollen** for bees.

Goldfinch
Thistle seeds

Thistles are wildflowers that produce seeds that are food for birds.

Earthworms live in the soil. Help them by not pouring poisonous weedkillers onto their home.

Soil
Earthworm

Did you know?

Worms help to recycle dead plants. They eat rotting bits of plants and then produce muddy worm poo. The poo adds lots of goodness to the soil and keeps it healthy.

Wildflower Wonders

Scatter some wildflower seeds in spring to make a flowery feeding patch for insects. If you only have a small outdoor space, use a flowerpot, bucket or even an old wheelbarrow.

You will need:
- A container
- Garden soil or peat-free potting compost
- A packet of mixed wildflower seeds
- A watering can

1. Ask an adult to make drainage holes in the bottom of the container.
2. Fill the container with soil or potting compost.
3. Sprinkle the seeds evenly over the compost. Lightly sprinkle another layer of compost on top.
4. Water well using the watering can.

Take pictures as your seeds start to grow. How long did it take for flowers to bloom?

Make Some Microhabitats

You can add mini habitats to your garden or school playground to provide homes for wildlife. These **microhabitats** will help a wide range of creatures.

Log pile

Make a pile of logs or a small, higgledy-piggledy pile of sticks. Spiders and beetles love this kind of woody habitat.

Did you know?
If there are lots of yummy minibeasts to eat, birds will visit to snap them up!

This starling is looking for tasty insects and spiders to eat in a rotting log.

Twigs and sticks microhabitat

A small pile of rocks or pebbles is a perfect shelter for insects, woodlice, snails and slugs.

12

Sweep or rake up autumn leaves. Make them into a messy pile in a quiet, cool corner of your outdoor space. This provides a warm, winter home for hedgehogs, mice, frogs, toads and other wildlife.

Mouse

Top Tip
Your leaf pile will start to rot. Once the leaves are dark, soft and crumbly, you can sprinkle them on the soil as food for worms.

Mini Wildlife Garden

If you only have a little outdoor space, make mini habitats in an old plant container. Position your mini garden in a quiet corner and watch for signs of wildlife.

You will need:
- An old plant container
- Garden soil or peat-free potting compost
- A yoghurt pot
- Twigs, bark or a log
- Different-sized stones
- Leaves

1. Fill the container with soil.

2. Bury the yoghurt pot so it's level with the soil and fill it with water.

3. Put a twig in the water. If a small animal falls in, it can use the twig like a ladder to escape.

4. Create your habitats! Add twigs, a log, stones and leaves. Weeds will soon sprout in the soil, too.

Keep a diary of the wildlife that visits your mini garden.

13

Minibeasts Are Cool...

...and Damp

When you visit a woodland take a look down. The shade of the trees makes the ground damp. The soil is covered with rotting leaves, **moss**, bark and ferns.

Many minibeasts and amphibians need cool, damp places to live.

Moss

Bark

Top Tip
Next time you go for a woodland walk, gather some moss, sticks and bark in a reusable bag to bring home.

Make a Flowerpot Burrow

A flowerpot makes a great burrow or hiding place for a toad or frog.

1. Choose a shady spot under ferns or shrubs, away from direct sun.
2. Dig a hole big enough for half the pot to be buried in.
3. Fill half of the pot with soil.
4. Place the moss, stones and grass inside. Wait for a frog or toad to move in!

You will need:
- A spade
- A flowerpot
- Moss
- Stones
- Long grass

Toad

Did you know?
A toad chooses a damp, dark, secret place to be its burrow. It rests in its home all day. At night, it hunts for minibeasts.

Place leaves, moss, stones, bark and sticks in a cool, dark corner of your garden or school playground.

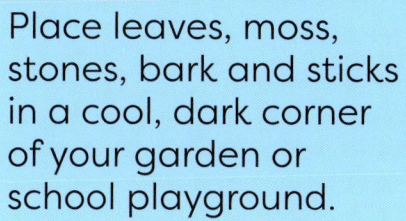
Centipede

Centipedes are fierce hunters that eat insects, spiders and woodlice.

Woodlouse

A damp habitat will attract woodlice, centipedes, earwigs, millipedes and other COOL minibeasts.

Did you know?

Millions of years ago, woodlice lived in water. Over time, they adapted to live on land. However, they still need water to breathe. They have body parts that take oxygen out of tiny amounts of water.

Slugs and snails prefer damp habitats. That's because their slimy bodies must stay cool and moist.

Snail

Slug

Slugs and snails provide food for frogs, toads and birds.

15

Beautiful Beetles

Beetles recycle rotting wood and help new plants to grow. And they are food for birds, frogs, toads, bats, squirrels and mice.

Many kinds of beetles eat dusty pollen, so let's grow flowers for beetles.

Yellow pollen

Beetle

Did you know?
When beetles visit flowers, they spread pollen from flower to flower on their bodies. This helps flowers produce seeds.

Male stag beetle

Female stag beetle

Stag beetle larva eating wood

If you have space, partly bury some small logs in the soil. Stag beetles lay their eggs in the soil beneath tree stumps and logs. When **larvae** hatch from the eggs, they feed on the rotting wood.

Stuff flowerpots with straw and dried plant stems, then hang them in your garden. They make great hiding places and warm winter nests for ladybirds and other insects.

Ladybird

Did you know?

Ladybirds and many other types of beetles **hibernate** in winter. During colder months, they hide in warm, safe places and sleep until spring.

Build a Bug Hotel

Make this bug hotel to give minibeasts a cosy home. If you want to paint the tins, make sure you do this first and let them dry. Gather the dry materials you need from your garden, or when you go for a nature walk.

You will need:
- Clean, recycled tins with no sharp edges
- Superglue
- Dry leaves, dry grass, twigs, pinecones, bamboo canes, bark and moss
- Thick string

1. Carefully stick the tins together using the superglue. You can use as many tins as you like, we suggest 6 or 7. Leave to dry.

2. Once the tins are firmly glued together, stuff them with your dry materials. Try a mix of materials in each tin.

Bamboo cane

If you want to include bamboo canes, ask an adult to cut or saw the canes into pieces the length of your cans.

3. Place your bug hotel in a quiet spot in your outdoor space. Or tie string around the cans and hang them on an outside wall.

4. Wait for your buggy guests to move in.

17

Create a Buzz!

Bees need flowers to get nectar and pollen to eat. There are lots of things we can do to keep bees buzzing.

Did you know?

Many foods we eat, including tomatoes and apples, come from plants that need bees. Bees carry pollen from flower to flower on their bodies. This helps tomato plant flowers and apple tree blossoms produce seeds and new fruits.

Bees

Sunflower

Sunflowers are easy to grow from seeds and are a perfect bee food.

If you don't have a garden, grow flowers in pots or window boxes. Bees will soon find you!

Let **clover** grow in the lawn. Bees love to feed on clover.

Clover

If you find a tired bee on the ground, you can help. Dissolve some sugar in warm water and give the bee a drink.

Bottle cap of sugar water

Top Tip
Bees rarely sting. When you see them, just let them BEE. Tell your friends about the good work bees are doing.

Did you know?
Wasps help to spread pollen, too. If you see a wasp, stay calm and quiet. There's no need to harm it.

Make a Buzzy Bee Planter

Make this fun, recycled planter and fill it with bee-friendly flowers.

You will need:
- A clean, empty plastic jar with a lid
- Masking tape
- Scissors
- Yellow, black, white and red acrylic paint
- Paintbrushes
- An adult helper with a craft knife
- Peat-free potting compost
- Bee-friendly plants
- String

1. Stick three strips of masking tape at evenly spaced intervals around the plastic jar.

2. Paint the entire jar yellow, painting over the masking tape. Leave to dry.

3. Paint a bee's face on the lid of the jar. Leave to dry.

4. Remove the strips of masking tape from the dry jar. Paint the clear sections black, to create bee stripes. Leave to dry.

5. Ask your adult helper to cut out a rectangle shape from one side of the jar. Poke holes in each side and thread through some string.

6. Fill the bee jar with compost. Add your bee-friendly plants, and hang your buzzy bee planter outside.

19

Create a Butterfly Garden

Butterflies need our gardens and other outdoor spaces. That's because their wildflower meadow habitats are disappearing.

If you have a grassy area, let dandelions and daisies grow for butterflies to feed from.

Swallowtail butterfly

GROW HERBS IN POTS

Place herbs in a sunny spot, and butterflies will settle to feed on the flowers.

PLANT A BUDDLEIA

Peacock butterfly

Its flowers are packed with nectar.

Did you know?

As butterflies flutter from flower to flower, they spread pollen, and help new plants grow.

MAKE A MUD PUDDLE

Butterflies drink the gloopy mud, which contains goodness they need.

20

Top Tip
Let a patch of stinging nettles grow. Many types of butterflies lay their eggs on them. When the caterpillars hatch, they eat the nettles.

Peacock butterfly caterpillar

Make a Butterfly Sponge Feeder

Butterflies love a sugary drink. Make these butterfly-shaped feeders soaked in sugar water.

You will need:
- New, compostable kitchen sponges
- A jug
- 2 cups of boiling water
- 1 cup of sugar
- A small spray bottle
- String
- Scissors
- An adult helper

1. Wash the sponges in cold water to soften them.

2. Ask your adult helper to pour two cups of boiling water into the jug. Carefully, add the sugar and let it dissolve.

3. Allow the sugar water to cool completely then pour it into the spray bottle.

4. Pinch a sponge to make a butterfly shape, then tie it with the string. Do the same to the other sponges.

5. Hang your sponge butterflies outside. Spray them all over with the sugar water.

Sit back and watch the butterfly ball!

Let's Feed the Birds

You can help care for garden birds by providing them with food. As soon as you start putting food out, hungry birds will arrive!

You can feed the birds all year round, but especially in winter and spring.

Bird feeder

Did you know?

In the cold winter months, there are fewer insects and seeds for birds to eat. In spring, parent birds are busy raising young. They will be grateful for a filling snack!

Fruits such as apples and pears are good choices. Place them on bird tables or wooden posts.

Top Tip

Always put bird food in a place that's high up, so cats can't pounce on the feeding birds.

Father blackbird

Worms

Chick

Encourage butterflies into your outdoor space – their caterpillars are food for birds.

Put out water for birds, too. They will drink and splash around, having a bath.

Keep soil healthy and weedkiller-free for earthworms. Many birds eat worms and feed them to their chicks.

Old frying pan

Make a Bird Café

Add a café (feeding station) to your garden and watch the bird customers flock in!

You will need:
- A bird table or place for hanging feeders
- Bird feeders
- Birdseed, peanuts, suet balls, mealworms, raisins or sultanas, chunks of fruit

1. Position your café in a safe place away from cats and dogs.

Ask for feeders for your bird café as a birthday present!

2. Remove any nylon mesh bags from shop-bought bird food. Place the food in the feeders and on the table.

Suet balls

Seeds

Peanuts

3. Watch from a distance. How many different birds can you name?

Best for Nests

In the spring, birds build nests in which to lay their eggs.

Try creating some safe nesting areas and putting out nesting materials. This will help birds to build warm, cosy nests for their chicks.

Nesting material

Did you know?
Birds need twigs, grass, moss, animal hair and sometimes mud to build and line their nests.

Do you have somewhere to put a nest box? A quiet, high spot on a garden fence or a wall is a good place.

Top Tip
A tray or box can be used to offer your bird neighbours a selection of nesting material. Dry leaves, moss, grass and even pet hair are good materials.

Fluffy seedheads
Dry grass
Free Nest Material

Some birds like to build nests in leafy, secret places. If you have hedges or bushes, let them grow undisturbed to keep nests hidden.

Did you know?

The best time to provide nesting material is March to July. This is when birds are at their busiest setting up home.

Old Boot Bird Nest

Recycle a pair of old boots and provide a safe place for a bird to nest.

You will need:
- Old boots, such as hiking boots or welly boots
- Dry grass, moss and other soft, natural nesting materials
- String
- An adult helper with a drill and screws

1. Find a safe, quiet place to put a boot bird nest. For example, on the outside of a shed or in a tree.

2. Stuff the boot with nesting material.

3. Use string to tie the boot in a tree or bush. Or ask an adult to fix the boot to the outside of a shed or porch.

4. Wait for a family of birds to move in!

Make a Splash for Water-Lovers

All wildlife needs water. Frogs, toads and some insects lay their eggs in water. Birds and mammals need fresh water to drink.

Make a Mini Pond

Create a mini pond in a sheltered spot amongst plants and flowers. This will give wildlife a safe place to live, drink and find food.

You will need:
- A spade
- Gloves
- A container such as a plastic washing-up bowl or shallow storage box
- Rocks and pebbles
- A log
- Pondweed

1. Dig a hole in your chosen spot. It should be the size of your container and almost as deep. Put your container in the hole.

2. Fill the sides of the hole with soil so the container is snug.

3. Add rocks and pebbles to the container, to act as stepping stones.

4. Place a log in the pond, so it slopes. This is an escape ramp for small animals.

5. Fill the pond with water, but not to the top. Allow the rocks and log to break the water's surface.

6. Add some pondweed such as hornwort and water milfoil. Your pond is ready for wildlife!

You can put rocks around the edge of the pond, too.

Frog eggs
Toad
String of toad eggs

In early spring, frogs lay clumps of eggs in ponds, while toads lay strings. Check for eggs in your mini pond.

A tadpole forms and wriggles out of the jelly.

Jelly
Egg
Tail
Tadpole

An adult frog or toad lays eggs.

The young amphibian leaves the pond and grows into an adult.

A tadpole grows front legs and becomes a froglet or toadlet.

A tadpole grows back legs and its tail shrinks.

Did you know?

Frogs and toads go through an amazing lifecycle.

An emperor dragonfly laying eggs

Look out for dragonflies darting over your pond as they hunt for food. A dragonfly may also lay her eggs in your mini pond.

27

Keep Caring for Wildlife

Every small action you take to help wildlife makes a difference!

HELP HABITATS
Keep wild places clean and safe for wildlife by picking up litter.

PLANT FOR WILDLIFE

By growing plants that insects love, you are helping all kinds of animals, from birds to bees. Wildlife needs wildlife!

Top Tip
Caring for wildlife is good for everyone. Share this book with your friends and ask them to join in the activities with you.

USE YOUR OUTDOOR SPACE
Let things grow wild and add wildlife-friendly corners and mini habitats.

Make a Wildlife Planner

Design and draw a planner that will help you care for wildlife throughout the year. Use the activities in this book to get you started.

You will need:
- A piece of A3 card
- Colour pencils
- A ruler

1. Using the ruler, divide the card into 3 columns with 4 panels in each column, one panel for each month of the year.

2. Colour in your panels and add wildlife designs, such as bees, butterflies and flowers.

JANUARY	FEBRUARY	MARCH
Feed birds throughout the year, but especially in winter and spring.	Make a boot bird nest.	Sow wildflower seeds.
APRIL	MAY	JUNE
Make a mini wildlife garden.	Make a buzzy bee planter.	Create a butterfly garden.
JULY	AUGUST	SEPTEMBER
Make a butterfly sponge feeder.	Build a mini pond.	Build a bug hotel.
OCTOBER	NOVEMBER	DECEMBER
Make a flowerpot home.	Plant a tree.	Feed the birds.

3. Write the name of a month in each panel.

4. Finally, add the activities from this book. Can you think of more things you can do each month? Stick your planner to a wall.

Even if you don't have a garden, you can still create small habitats, feed the birds and help bees and butterflies. Remember – every action counts!

29

GLOSSARY

amphibian
An animal group that includes frogs, toads and newts. Amphibians lay eggs in water. Young amphibians, such as frog and toad tadpoles, live in water. Most adult amphibians live on land and in water.

clover
Small, low-growing plants that have white or pink flowers. Clover helps to keep soil healthy and provides insects with pollen and nectar.

habitat
A place where plants, animals and other living things make their home. Woodlands, gardens and ponds are all types of habitats.

hibernate
To spend winter in a deep sleep without eating or drinking.

insect
A small animal with six legs and a body in three main parts. Most insects have wings. Bees, wasps and butterflies are all insects.

larva
A young insect. A larva hatches from an egg. Caterpillars and beetle grubs are both types of larvae.

mammal
A warm-blooded animal with fur or hair on its body. Mammals give birth to live young and feed them milk. Mice, squirrels and foxes are all types of mammals.

microhabitat
A very small habitat. The word "micro" means tiny.

moss
A tiny plant that often grows on rotting logs or rocks. Many moss plants grow close together and look like a soft, green carpet.

nectar
A sweet, sugary liquid that's produced by flowers. Insects such as bees and butterflies feed on nectar.

pollen
A fine, powdery dust that's produced by flowers. Pollen helps flowers produce seeds and is a food for beetles, bees and many other animals.

weed
Another name for a wild plant. Weeds are often tough and quick-growing, but they produce food for many insects and other wild animals.

INDEX

A
amphibians 4, 7, 10, 13, 14–15, 16, 26–27

B
bees 10, 18–19, 28–29
beetles 12, 16–17
birds 4–5, 6, 11, 12, 15, 16, 22–23, 24–25, 26, 28–29
Build a Bug Hotel 17
butterflies 7, 20–21, 23, 29

C
centipedes 15

D
dragonflies 27

E
earthworms 11, 13, 23
eggs 7, 16, 21, 24, 26–27

F
flowers 7, 10–11, 16, 18–19, 20, 26, 29
frogs 7, 10, 13, 14–15, 16, 26–27

H
habitats 4–5, 6–7, 8, 10–11, 12–13, 14–15, 16–17, 20–21, 26–27, 28–29
hibernation 17

I
insects 4–5, 7, 11, 12, 15, 16–17, 18–19, 20–21, 22–23, 26, 28–29

L
litter 4–5, 6–7, 8–9, 28
Litter Pick and Sort 9

M
Make a Bird Café 23
Make a Butterfly Sponge Feeder 21
Make a Buzzy Bee Planter 19
Make a Flowerpot Burrow 14
Make a Mini Pond 26
Make a Wildlife Planner 29
mammals 4, 6, 8, 13, 26
mice 6, 13, 16
Mini Wildlife Garden 13

N
nectar 7, 10, 18, 20

O
Old Boot Bird Nest 25

P
Plant a Tree 7
pollen 10, 16, 18–19, 20
ponds 7, 26–27, 29

S
seeds 7, 11, 16, 18, 22–23, 24, 29
slugs 12, 15
snails 12, 15

T
toads 13, 14–15, 16, 26–27

W
water 5, 13, 15, 19, 21, 23, 26–27
weedkiller 10–11, 23
weeds 10, 13
Wildflower Wonders 11
Wildlife Water Bowl 5
woodlice 12, 15